Oxford International Primary Science

Workbook

3

Terry Hudson
Debbie Roberts

OXFORD
UNIVERSITY PRESS

OXFORD
UNIVERSITY PRESS

Great Clarendon Street, Oxford, OX2 6DP, United Kingdom

Oxford University Press is a department of the University of Oxford. It furthers the University's objective of excellence in research, scholarship, and education by publishing worldwide. Oxford is a registered trade mark of Oxford University Press in the UK and in certain other countries

British Library Cataloguing in Publication Data
Data available

9780198376446

13

Paper used in the production of this book is a natural, recyclable product made from wood grown in sustainable forests. The manufacturing process conforms to the environmental regulations of the country of origin.

Printed in Great Britain by Ashford Colour Press Ltd.

Acknowledgements
The publishers would like to thank the following for permissions to use their photographs:

Cover: Frans Lanting/Corbis; **p43**: thirboy/iStock; **p70**: Shutterstock; **p71l**: CHEN WS/Shutterstock.com; **p71r**: Bob Daemmrich/Alamy Stock Photo.

Although we have made every effort to trace and contact all copyright holders before publication this has not been possible in all cases. If notified, the publisher will rectify any errors or omissions at the earliest opportunity.

Contents

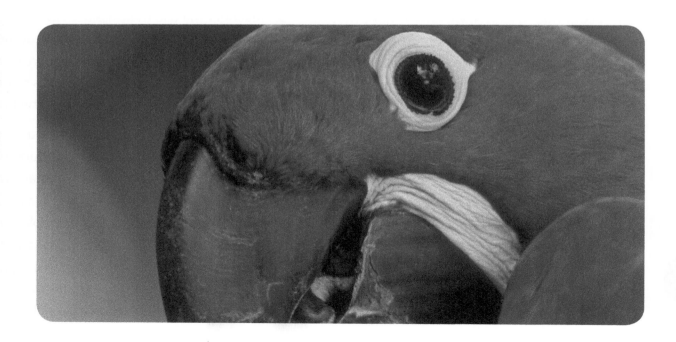

How to use this book

This workbook has been written to support the Student Book that students are using at school. The Student Book has some write-in and hands-on tasks to help students learn and to test their understanding, but it is important to extend these tasks – including home learning.

This workbook is split into the six modules of the Cambridge curriculum for this stage. These are:

1 Life Processes

2 Materials

3 Flowering Plants

4 Introducing Forces

5 The Senses

6 Keeping Healthy

Each module begins with extra support to enable you to help the student. It explains the focus of the module and then provides specific information and advice about how the student can be supported. We all learn new skills and knowledge initially with lots of support but we cannot continue to learn if we always have that support. In other words, allow the student to try things on their own or with some guidance, and only step in if the student shows genuine confusion or frustration.

The activities build on the work at school and are aimed at developing language skills, scientific enquiry skills and understanding rather than just recall.

Each module ends with a review task that the student is asked to do at home. This review and reflection is a key aspect of learning.

Why is home learning important?

Encouraging students to think about and apply their growing skills and knowledge outside the classroom, and especially at home, allows them to consolidate understanding and to practise activities. This helps with confidence. They also have opportunities to see that science is relevant all around them – not only during science activities in school. Another advantage of home learning is that you can find out what the student is studying and show your genuine interest. The student may even be able to teach you some science. Finally, home learning can be fun and help the student to develop good learning and study habits that will help them throughout their life.

Class activities and home learning activities

Each activity in this workbook has an icon to give the student, you and the student's teachers a clear idea of the nature of the task. These icons are explained on page 5.

Each module has four **class activities** and eight **home learning activities.**

- Teachers will use the class activities to supplement the Student Book and additional resources. The class activities involve group discussion with peers and activities requiring equipment not easily obtained in the home.

- The home learning activities cover a range of different activity types that encourage the development of science skills. In many of the home learning tasks the student is asked to involve people at home. For example, the student asks people at home to take part in team games, investigations and surveys, and gives a presentation explaining what the student has learned about materials.

Activities and icons

 Writing activities

There are spaces for students to record short answers to key questions using the information on the pages and from their prior learning in school. Sometimes a drawing is required.

 Discussion activities

Students are encouraged to discuss scientific ideas and approaches and are expected to work in pairs and small groups for this type of work.

 Investigations

Students are encouraged to record plans, scientific notes and results for each investigation. They are asked to make predictions and to compare their results with others.

 Measuring activities

Students are given opportunities to use and develop a number of key measuring skills. There are step-by-step instructions and advice about how to use measuring equipment accurately.

 'Think about' questions

Students are encouraged to consider scientific phenomena and to try to explain them using the knowledge and understanding they have gained during the topic. These questions may also involve thinking of examples from students' own experience.

 Extra support

This icon indicates where the adult at home can find advice on how to help the student with each home learning activity.

1 Life Processes

Extra support

Introduction

The main purpose of this module is to help students to understand more about living things. Students are first encouraged to use the physical appearance of objects to sort them into groups.

Students will also learn to identify and name some different types of vertebrates (animals with a backbone). Next, students are asked to think about the essential life processes of animals and what animals need in order to stay alive, including the foods different animals eat.

Students then learn that animals use their senses to stay alive and that animals and plants have a life cycle that involves growing up, maturing and reproducing. Students look at differences between living and non-living things, and between plants and animals. They consider what a plant needs in order to grow well.

This module will help students to practise these scientific enquiry skills:

- observation – collecting evidence by looking and measuring (pages 8, 9, 10, 11, 12, 14, 16, 17, 18)
- planning – asking questions and planning how to seek answers (pages 9, 11, 12, 18)
- predicting – stating what they think will happen and then comparing this with what actually happens (page 18)
- recording – writing or drawing observations or stages in work (pages 8, 9, 11, 13, 14, 16, 18)
- making comparisons – comparing sets of evidence or data (pages 8, 9, 11, 12, 13, 14, 16, 17, 18).

Ways to help

Encourage the student to identify groups of objects, plants and animals and to use simple keys to help them. Remind them that keys often work on a yes–no basis. You could talk about the differences between living and non-living things and use examples in the home. Help the student to appreciate that there are many different types of vertebrates, such as mammals, reptiles and amphibians, but that they all share the key characteristic of having a backbone. Allow the student to see pictures of mammals that have adapted to live in the sea or fly, to show that there are exceptions to general rules about which animals belong in which groups. Also help the student to measure and estimate sizes of living things to help them understand how animals and plants grow. You can also show them examples of young animals and their parents for comparison.

Remember to encourage and help the student but try to let them find out as much as they can on their own.

Finally, help the students complete the 'What I have learned…' summary to test their understanding and recall.

Key words

animal	non-living
breathe	plant
dark	reproduce
feed	see
food	seeds
grow	senses
hear/hearing	sight
human	smell
light	taste
living	touch
move	water

Scientific enquiry words

compare	look
group	name
investigate	sort

 Helping with activities

The following guidance is intended to offer advice to the parent, or other adult at home, on how to help the student with each home learning activity.

Types of vertebrates (page 10)

The questions encourage the student to think about the characteristics of vertebrates. Whales live in the sea and bats can fly, but both are mammals. They have warm blood and feed their young with milk. They also breathe through lungs. Despite not being able to fly, the penguin is a bird so it lays eggs and is covered with feathers.

Finding fruit and vegetables (page 11)

This activity helps the student to structure a survey. It would be helpful to take them to a nearby shop or market to see a wide variety of fruits and vegetables.

The hearing game (page 12)

The purpose of the game is to explore hearing. You may not be surprised by how good most people are at detecting the direction of sounds, but it may be a surprise to find out how easily our hearing can be tricked – especially with hands acting as ears that point backwards.

Your family timeline (page 14)

Find a long piece of string to stretch across a room; alternatively, spread the pieces of paper along a wall or the floor. The student can fold sticky notes over the line, stick them to a wall, or attach small pieces of paper to the string using paperclips.

Find the important words (page 15)

The wordsearch will help the student to practise the key words relating to care of offspring. Encourage the student to start by looking for beginning letters or combinations of letters.

How animals move (page 16)

Watch the student as they think of examples of animals that move in different ways. The student should be able to think of two more things that living things can do from: breathe, eat, drink, grow and reproduce.

How are animals and plants different? (page 17)

This activity helps the student to understand that, although both living, plants and animals are very different. One possible misconception is thinking that plants move from place to place: (they cannot do this although they do move towards light and they do grow).

What I have learned…. (page 19)

Help the student to complete the crossword to review their learning from the module.

What group does this animal belong to?

See Student Book 3, pages 6–7

Class activity Sorting objects

You have learned how we **sort animals** into **groups** (see your Student Book pages 6–7). In this activity you will be sorting **non-living** objects. What an object looks and feels like is known as its physical appearance.

We can use the physical appearance of objects to **sort** them into **groups**.

Look at the objects in the picture.

Sort the objects into three **groups**.

Draw the objects in the circles. Use one circle for each **group.**

Think of a **name** for each **group**. Write the **name** at the top of the circle.

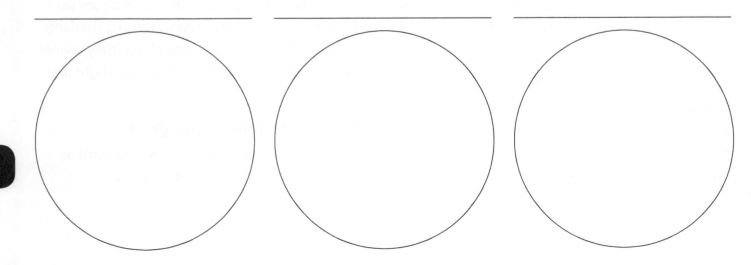

What group does this animal belong to?

See Student Book 3, pages 8–9

Class activity Using a key to identify animals

A simple way to **group** and identify animals is to use a key. A key helps us to get 'yes' and 'no' answers.

Look at the key. It will help you to identify some invertebrates.

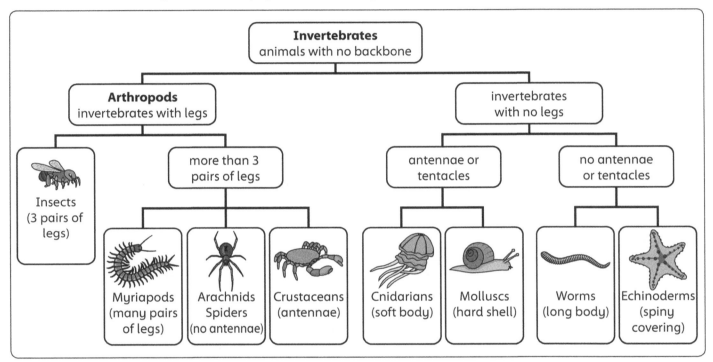

Invertebrates
animals with no backbone

Arthropods
invertebrates with legs

invertebrates
with no legs

more than 3
pairs of legs

antennae or
tentacles

no antennae
or tentacles

Insects
(3 pairs of
legs)

Myriapods
(many pairs
of legs)

Arachnids
Spiders
(no antennae)

Crustaceans
(antennae)

Cnidarians
(soft body)

Molluscs
(hard shell)

Worms
(long body)

Echinoderms
(spiny
covering)

Use the key to **name** the animal types in the pictures.

_____   _____   _____

Write two reasons why the spider is not in the same part of the key as the snail.

1 _____

2 _____

What group does this animal belong to?

See Student Book 3, pages 10–11

Home learning Types of vertebrates

If you need help to answer these questions, use books or the Internet.

For advice on how to help the student with this activity, see page 7.

Why is a whale a mammal and not a fish?

Write two reasons.

1 _____

2 _____

Why is a penguin a bird and not a fish?

Write two reasons.

1 _____

2 _____

3 Why is a bat a mammal and not a bird?

Write two reasons.

1 _____

2 _____

GOING FURTHER

Try to find a picture of a duck-billed platypus.

What **group** of animals does this animal belong to?

_____

Staying alive

See Student Book 3, pages 12–13

Home learning Finding fruit and vegetables

✋ **Fruit and vegetables** (see page 13 of your
Student Book.)

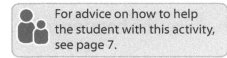

For advice on how to help
the student with this activity,
see page 7.

✏️ Answer the questions to help you make your survey of fruit and
vegetables scientific.

1 How many times will you carry out your survey? _____

2 Would the results be different if you did the survey in January or in July?

_____

3 How will you record (write) your results?

_____

4 How will you **compare** your results with other people's results?

_____

5 How will you present your results? Will you use a table or a graph or

both? _____

6 Which of these graphs might be a useful way to show people your
results? Tick ✓ your choices.

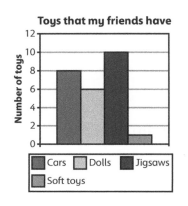

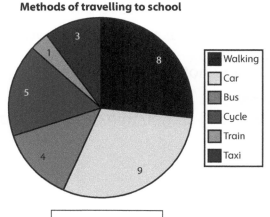

a bar chart

b line graph

c pie chart

Staying alive

See Student Book 3, pages 14–15

Home learning The hearing game

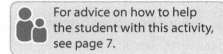

For advice on how to help the student with this activity, see page 7.

Play the **hearing** game.

Play this game with family and friends to **investigate** hearing.

1 Ask everyone except one person to sit in a circle.

2 Ask the remaining person to sit in the centre of the circle.

3 Blindfold the person in the centre if they do not mind. If they do mind, ask them to close their eyes.

4 Point to a person in the circle. They clap their hands once loudly.

5 The blindfolded person must point in the direction of the sound.

6 Point to different people, so the blindfolded person **hears** the clapping sound in front of them, at the side and behind them.

7 Ask the blindfolded person to cup their hands behind their ears. Do the investigation again.

8 Ask the blindfolded person to cup their hands in front of their ears with their hands pointing backwards. Do the investigation again.

Do you notice any differences?

Why do you think some animals turn their ears to listen more carefully?

Staying alive

See Student Book 3, pages 16–17

Class activity How tall are animals?

✋ **Compare** the heights of different animals.

✏️ Use the data in the table to draw a bar chart of the heights of the animals.

Animal	Height (cm)
goat	60
horse	200
human	170
rabbit	20
cat	40

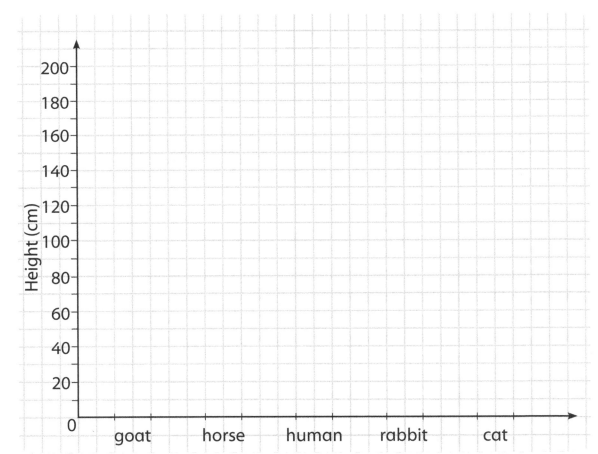

✏️ Which animal is the tallest? _____

Which animal is the shortest? _____

Is it easier to **compare** heights using the table or the bar chart? Why?

_____

Home learning Your family timeline

 Make a family timeline.

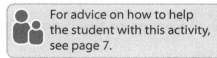
For advice on how to help the student with this activity, see page 7.

You will need: a long piece of string, some pieces of paper and some paperclips.

Divide your string into six sections. Each section covers a range of ages.

Section 1: 0–10 years old

Section 4: 31–40 years old

Section 2: 11–20 years old

Section 5: 41–50 years old

Section 3: 21–30 years old

Section 6: over 51 years old

Find out the name, age and height of as many of your family and friends as you can. Try to include at least eight people.

Write the name, age and height of each person on a separate piece of paper.

Use your string to make a timeline. Clip the paper with details of the youngest person at one end of the timeline. Clip the paper with details of the oldest person at the other end. Add the other pieces of paper in order.

Look at your timeline. Is there a link between the age of the people and their heights? Can you **see** a pattern?

_____

_____

Home learning Find the important words

Can you find all of these words in the wordsearch? Circle each word when you find it.

 For advice on how to help the student with this activity, see page 7.

offspring	reproduce	eggs
species	care	babies

d	o	v	s	p	e	c	i	e	s
i	s	o	l	c	t	f	n	c	y
m	e	g	g	s	p	g	k	b	a
o	f	f	s	p	r	i	n	g	x
v	y	p	g	u	c	a	r	e	v
h	r	e	p	r	o	d	u	c	e
q	v	y	t	n	h	a	l	j	t
a	z	b	a	b	i	e	s	p	u

When you find each word, **discuss** with an adult at home what you think it means.

Write three ways that adult **humans** care for their babies.

1 _____

2 _____

3 _____

Is this living or non-living?

See Student Book 3, pages 20–21

Home learning How animals move

The table shows some of the ways that animals can **move**.

Think of an animal that moves in each way. Write the **name** of the animal in the table.

For advice on how to help the student with this activity, see page 7.

Can humans move in this way too? Write the answers in the table.

Way of moving	An animal that does this	Can humans move in this way? Yes or no?
Run		
Walk		
Jump		
Swim		
Crawl		
Slide		
Fly		

One way we know that animals are **living** things is because they can move.

Write two other things living things can do.

1 _____

2 _____

Write two non-living things you have seen today.

1 _____

2 _____

How can we help plants to grow?

See Student Book 3, pages 22–23

Home learning How are animals and plants different?

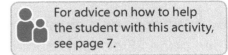

For advice on how to help the student with this activity, see page 7.

Look at the picture of the **plant** and person. Can you match each label to the correct picture? One label can be matched to both pictures.

I can move from place to place.

I **grow** and produce **seeds**.

I can make my own **food** using energy from the sun.

I need **water**.

I need to eat food.

Use a coloured pencil and a ruler to draw a straight line from each label to the correct picture.

Which label did you link to both the plant and the person?

_____

Write two ways that plants and **humans** are different.

1 _____

2 _____

Class activity Planning a fair test

What does a plant need to grow well?
(See page 25 of your Student Book.)

Your teacher may have given you a template to help you plan your investigation. This helps you to think through your investigation in a scientific way. It helps you to plan a fair test.

These questions will help you to think like a scientist about your investigation.

1 What were you trying to find out?

2 What did you predict would happen?

3 How did you set up your investigation?

- What equipment did you need?

- What factors did you keep the same?

- What factors did you change?

Draw your plants to show how they looked after four weeks.

Plant A – Sun and water	Plant B – In cold and **light**
Plant C – No water	Plant D – In the **dark**

Home learning What I have learned...

Solve the clues and complete the crossword puzzle. This will review your understanding of this module.

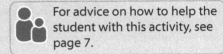

For advice on how to help the student with this activity, see page 7.

When you have completed the puzzle, you will see an extra word in the shaded squares.

Clues

1 This **group** of animals have feathers and wings.

2 Animals need to do this to stay alive. It is another way of saying 'to eat food'.

3 These living things use energy from the Sun to make their own food.

4 This **group** of animals have scaly skins and lay eggs.

5 Animals can do this to find food or escape from danger.

6 Snakes do this to move from one place to another.

7 We use this **sense** to feel whether objects are rough or smooth.

8 This building material is non-living.

9 We have five of these: **sight**, hearing, **smell**, **touch** and **taste**.

Write the word shown in the shaded squares. _____

What does this word mean? _____

2 Materials

Extra support

Introduction

This module helps students to understand the different materials around them. They will learn that every material has physical properties and that these are linked to the uses of a material. They will distinguish between magnetic and non-magnetic materials and explore the uses of different materials. Students will investigate the insulating properties and absorbency of materials. They will also learn that we can group similar materials together according to their type and their properties.

This module will help students to practise these scientific enquiry skills:

- observation – collecting evidence by looking and measuring (pages 22, 23, 25, 26, 27, 28, 29, 30, 32)

- planning – asking questions and planning how to seek answers (pages 25, 26, 27, 29, 30)

- predicting – stating what they think will happen and then comparing this with what actually happens (pages 25, 26, 29, 31)

- recording – writing or drawing observations or stages in work (pages 22, 23, 25, 26, 27, 28, 29, 30, 31, 32)

- making comparisons – comparing sets of evidence or data (pages 22, 25, 26, 27, 29, 30)

- drawing conclusions – examining results to identify any patterns and/or to suggest explanations (pages 25, 27, 29).

Ways to help

Encourage the student to apply their knowledge of the properties of materials to identify objects. Discuss with the student the link between the properties of a material and its uses. Point out common object such as pans and tools and ask the student to imagine them being made of another, less suitable material. Point out materials with different properties, such as soft, hard, strong and magnetic. Allow the student to carry out home surveys to find objects and to test and identify the materials. For example, help them to find magnets in the home, such as fridge magnets, and catches on doors, handbags and toys, and to consider what they are used for.

Encourage the student to think scientifically when they explore properties, such as slippery surfaces. Help them to develop the skills of predicting, planning and testing in order to obtain results and draw conclusions.

Remember to encourage and help the student but try to let them find out as much as they can on their own.

Finally, help the student complete the 'What I have learned …' summary activity to test their understanding and recall.

Key words

absorbent	object
attracted	rough
hard	see-through
magnet	shiny
magnetic	smooth
material	soft
non-magnetic	waterproof

Scientific enquiry words

decide	record
investigate	sort
predict	

 Helping with activities

The following guidance is intended to offer advice to the parent, or other adult at home, on how to help the student with each home learning activity.

Using properties at home (page 23)

Help the student to carry out a home survey to find objects that are hard, soft, shiny and see-through (transparent). You could set out some objects on a table or let the student find their own. Discuss any objects that have more than one property.

Materials game (page 24)

Each person says the name of an object. The other person has to say what they think is the best material, then the worst material, to make that object from. Think of the strangest examples of materials that you can, as this will emphasise that objects have to be made from suitable materials.

Slippery or safe? (page 25)

Help the student to use a homemade forcemeter. It works on the principle that the greater the force the more the elastic band will stretch. If there are no slippery floors in the home, include a smooth table top or kitchen work surface in the investigation.

Which objects at home are magnetic? (page 26)

The student will need a small magnet for this activity. A fridge magnet works well. Encourage them to predict whether or not different objects are magnetic. In fact, only iron, steel, cobalt and nickel objects are magnetic. It is important for the student to understand that metals such as copper and aluminium are non-magnetic, so try to include some objects made from these materials.

Making a compass (page 28)

Provide a sewing needle (or nail) and a magnet, as well as the other materials listed. A fridge magnet will work well. Make sure that the student strokes the needle with the magnet in one direction only in order to magnetise the needle. If you have a compass at home, help the student use this to determine North and South and perhaps the other compass directions. Help the student to use their compass to find a building to the North and West of their home.

Soaking up spills (page 30)

This activity is a test of the absorbency of different materials. The student will need four different materials, including some kitchen paper towels, and a measuring jug. Help the student to create a table to record their results.

Making a match (page 32)

Watch the student draw a line between each material and the correct set of properties. This is a good review of their understanding.

What I have learned... (page 33)

Allow the student to present their ideas about materials to a group of people at home. This is an excellent way for the student to show understanding of the module and to develop confidence and communication skills.

Materials around us

See Student Book 3, pages 30–31

Class activity Finding different objects

✋ Looking for properties

💭 Look around the room to find different **objects**.

Find objects that have the properties listed in the table. For example, you might find a rubber ball. You can write 'Rubber ball' next to 'Bouncy' because this is one of the ball's properties.

Property	Example of an object in the room
Hard	
See-through	
Soft	
Shiny	
Rough	
Smooth	
Breaks easily (brittle)	
Strong	
Heavy	
Bouncy	

✏️ Which properties were the easiest to find?

_____

Which properties were the most difficult to find?

_____

💬 Compare your table with another student's table. Did you find the same objects?

Home learning Using properties at home

Find and draw an object with each of these properties.

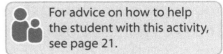

For advice on how to help the student with this activity, see page 21.

Find and draw an object that is **soft**.

Find and draw an object that is **shiny**.

Find and draw an object that is **hard**.

Find and draw an object that is **see-through**.

2 Materials

23

Materials around us

See Student Book 3, pages 32–33

Home learning Materials game

You might have played the **materials** game at school. This is a chance to play it with someone at home. You can teach them about materials.

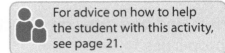

For advice on how to help the student with this activity, see page 21.

Ask someone to be your partner.

- Say the name of an object.

- Ask your partner to say what they think is the best material to make that object from. For example, you say 'bell' and your partner says 'metal'.

- Then ask your partner to say what they think is the worst material to make that object from. For example, they might say 'rubber' or 'glass'.

Now change roles. Let your partner choose an object. Try to think of some very useful materials and some useless materials for each object.

See Student Book 3, pages 34–35

Home learning Slippery or safe?

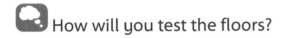 **Testing for slippery floors**

For advice on how to help the student with this activity, see page 21.

Plan and then carry out a test to **investigate** which floors at home are slippery and which are not.

How will you test the floors?

You can use a homemade forcemeter like the one in the picture.

Look at the forcemeter.

What will happen to the length of the elastic band if it is used to pull a heavy weight?

_____

What will happen to the length of the elastic band if it is used to pull a light weight?

_____

How can you use the forcemeter to measure how slippery a floor is?

In your test what will you keep the same? _____

What will you change? _____

How will you **record** and show your results? _____

Carry out your test.

After your investigation

Are there any floors in your home that are slippery? _____

Do you **predict** that hard floors are more or less slippery just after they have been washed? Circle your answer.

More slippery   Less slippery

Is it magnetic?

See Student Book 3, pages 36–37

Home learning Which objects at home are magnetic?

You will need a **magnet** for this activity. Your teacher might have given you one to take home. If not, you can use a fridge magnet.

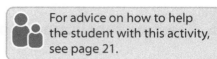

For advice on how to help the student with this activity, see page 21.

Which objects are **magnetic**?

Look around your home.

1 Choose six objects that you **predict** will be **attracted** towards a magnet.

Write the names of the objects into the table.

2 Choose six objects that you **predict** will not be attracted towards a magnet.

Write the names of the objects into the table.

3 Use a magnet to test your predictions.

The six objects I predict are magnetic	The six objects I predict are non-magnetic

How many of your predictions were correct? Tick ✓ all the objects you **predicted** correctly.

Magnets can be very useful at home.

Write three examples.

1 _____ 2 _____ 3 _____

Is it magnetic?

See Student Book 3, pages 36–37

Class activity Identifying N and S poles of a magnet

 How can we identify the poles of magnets?

You are going to see if you can identify the poles of magnets when they are covered up.

Your teacher has set up some magnets so you cannot see the poles. Any of the magnets could be arranged with the North pole towards you or away from you.

1 Your teacher will give you an uncovered magnet.

 How can you find out which pole is which on the covered magnets without taking the paper off?

2 Test each magnet in turn. **Record** your findings by writing 'N' or 'S' on the paper covering.

3 Your teacher will unwrap the magnets so you can check your answers.

 How many of the five did you get correct? _____

Which law of magnetism did you use to help you to work out your answers?

_____

Is it magnetic?

See Student Book 3, pages 38–39

Home learning Make a compass

 Make your own compass. The needle of a compass is a small magnet.

For advice on how to help the student with this activity, see page 21.

> You will need: a magnet, a sewing needle or nail, a black marker pen, a plastic bottle top, glue or sticky tape and a bowl of water.

1 Make the compass needle by stroking the magnet along a sewing needle or nail.

2 Always stroke in the same direction – not backwards and forwards. Stroke it about 12 times.

3 Mark the pointed end of the needle using a black marker pen.

4 Fix the needle to a plastic bottle top using glue or tape.

5 Float the bottle top and needle in a bowl of water.

6 When the needle stops spinning, notice which direction the needle points. This will be North–South.

7 If you have a compass at home, use it to find out if the black end of your needle is pointing North or South.

Once you know which directions are North and South, you can work out East and West.

> Take care with the needle. It has a sharp point.

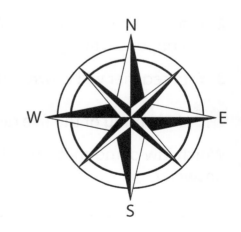

Using your compass

Use your compass to find:

■ a large or important building to the West of your home, which is: _____.

■ a large or important building to the North of your home, which is: _____.

Just right for the job

See Student Book 3, pages 40–41

Class activity Testing insulators

 Which material is best for a warm coat?

1 Set up the equipment shown in the picture.

> Remember: a good insulator stops heat from passing through it.

Predict which material will be the best insulator. _____

No insulator   Paper   Cotton   Felt

2 Add the same volume of water to each beaker.

3 As soon as you add the water to each beaker, **record** the temperature of the water.

4 Check the temperature of the water in each beaker every two minutes over a period of ten minutes.

5 **Record** your results in the table.

Time in minutes	Temperature (°C)			
	No insulator	Paper	Cotton	Felt
0				
2				
4				
6				
8				
10				

Which material is best for the warm coat? _____

GOING FURTHER

Thinking like a scientist

The materials were not all the same thickness. Was this a fair test?

If you put a lid on one of the beakers, will it be a fair test? Explain why.

2 Materials

29

Just right for the job

See Student Book 3, pages 42–43

Home learning Soaking up spills

Which material is the most **absorbent**?

If a material can soak up water, we say it is absorbent.

You can test to see how much water different materials absorb. Then you can compare them.

Test four different materials. One of these should be a paper towel.

For advice on how to help the student with this activity, see page 21.

1 Use a measuring jug to measure 200 cm^3 of water.

2 Put the material you are testing into the water for two minutes and then remove it.

3 Measure the volume of water left in the jug.

Design a table to **record** your results.

GOING FURTHER

Which test is better?

Compare the investigation you have just done to the one you did at school. Which one is more accurate? Explain why.

Just right for the job

See Student Book 3, pages 44–45

Class activity Invent an object

This activity will help you with the inventing task on page 44 of your Student Book.

✎ The name of our inventing team is _____.

The object we have chosen is _____.

Materials we might use are _____.

We have **decided** that the material we will use is _____.

The properties of this material are _____

_____.

💬 Think about these questions, so you can answer them when your class asks you.

■ Will your object be strong enough if it is made from this material?

■ Will it break easily?

■ Will it be too heavy?

■ Will it be too expensive?

■ Why is the object made from your material better than the original?

✎ Make some notes here.

_____

_____

_____

_____

_____

Home learning Making a match

Draw a line to link each material to its properties.

For advice on how to help the student with this activity, see page 21.

Metal

Fabric

Plastic

Glass

Hard, strong, shiny, easy to shape

Transparent, breaks easily, **waterproof**, shiny

Hard, strong, waterproof

Soft, bendy, absorbent

Write an example of how you use each material at home.

Metal _____

Fabric _____

Plastic _____

Glass _____

What we have learned about materials

See Student Book 3, pages 48–49

Home activity What I have learned...

You are going to give a presentation to teach people at home about materials. This will help you to review what you have learned in this module.

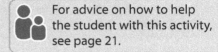
For advice on how to help the student with this activity, see page 21.

✏️ Use the clue boxes to plan your presentation. The questions will help you to think about what you already know.

Clue box 1: Properties

What are the properties of materials? _____
_____

Clue box 2: Common materials

What are the common materials? _____
_____

What are these materials used for? _____
_____

What can I say about magnetism? _____
_____

Clue box 3: Linking the property and the job

How will I explain that a material must have the correct properties to do its job?
_____
_____

Clue box 4: Sorting materials

How can I explain how to **sort** objects according to their properties?
_____
_____

3 Flowering Plants

Extra support

Introduction

During this module students will learn about flowering plants. This includes learning about the main parts of a flowering plant. Students will make a model of a flowering plant to help them understand the structure of a plant, and the function and importance of each part. Students will investigate what a plant needs in order to grow well. This includes investigating the effect of water and light on the health of the plant. Students will carry out a survey of their local area to identify how many trees are growing and where the trees are. Finally, they have an opportunity to make a greenhouse for a plant and test its efficiency.

This module will help students to practise these scientific enquiry skills:

- observation – collecting evidence by looking and/or measuring (pages 37, 39, 41, 42, 43, 44, 45)

- planning – asking questions and planning how to seek answers (pages 39, 41, 42, 45, 46)

- predicting – stating what they think will happen and then comparing this with what actually happens (pages 39, 42, 45)

- recording – writing or drawing observations or stages in work (pages 37, 39, 41, 42, 43, 44, 45)

- making comparisons – comparing sets of evidence or data (pages 37, 42, 44, 45)

- drawing conclusions – examining results to identify any patterns and/or to suggest explanations (pages 42, 45).

Ways to help

To support the student's learning about the parts of a flowering plant, help them to make and label models and to display diagrams at home. Carry out fun quizzes when you see flowers and plants. Growing bean seeds at home will allow the student to see root growth and you can also encourage them to find examples of roots forcing their way through roads and walls. Help the student to make a working model greenhouse for growing seeds, and encourage the student to investigate the impact of temperature on plant growth.

Use activities such as the scientific enquiry wordsearch and the 'photosynthesis' word game to help the student develop important language skills. Thinking about the meaning of the scientific enquiry key words they have been using at school will help the student develop the important skills needed to plan and carry out their own investigations.

Remember to encourage and help the student but try to let them find out as much as they can on their own.

Finally, help the students complete the 'What I have learned…' summary to test their understanding and recall.

Key words

compost	plant
dark	root
flower	stem
grow	temperature
healthy	unhealthy
leaf	water
light	

Scientific enquiry words

discuss	investigate
explain	measure
identify	observe

 Helping with activities

The following guidance is intended to offer advice to the parent, or other adult at home, on how to help the student with each home learning activity.

Main parts of flowering plants (page 37)

The student will bring home a model of a flowering plant that they have made at school. They will explain how their choice of materials relates to the functions of the plant parts. Create labels for each part of the plant and use them to label the student's model.

Scientific enquiry wordsearch (page 38)

As the student finds each word in the wordsearch, ask them what the word means. Ask them to use each word in an example sentence, providing help as necessary.

Photosynthesis competition (page 40)

Create two teams of people to take part in the competition. The teams make as many words as they can using the letters in the word 'photosynthesis'. Help the student to find two more words that begin with 'photo-', for example photograph, photocopy,

photogenic, photon. Discuss the words and point out that they are all words to do with light.

Grow a bean seed (page 41)

Help the student to grow a bean seed in a glass jar or other clear container. Encourage the student to observe the seed every day and to draw their observations every few days. Talk about how gravity pulls the roots down to the water and nutrients in the soil. The stem grows upwards to reach the light.

The power of roots (page 43)

Discuss with the student the fact that roots can be very strong, and even force their way through bricks and up through road surfaces. Take the student outside to a safe area to see examples. Encourage the student to draw what they observe and to record on their drawing where they saw the roots.

Tree survey (page 44)

Discuss the fact that trees are very big plants. Take the student outside to observe trees growing. If this is not possible, use the Internet, books or magazines to find pictures of trees found locally. Encourage the student to use a tally chart to record how many of each type of tree they find.

Make your own greenhouse (page 46)

Collect materials that the student can use to make their greenhouse: a container that will prevent water and soil spillage; cling film or clear plastic for the top; small sticks for holding up the top. Encourage the student to draw a design first. The student could use the greenhouse to grow seeds such as grass or cress.

What I have learned... (page 47)

This final activity reviews the student's learning from the module and provides practice in answering questions that could be used in assessments.

Class activity Make a model of a plant

✏️ Label the **roots**, **stem** and a **leaf** and **flower** on the picture.

✋ Make a model of a **plant**.

1   Your teacher will give you a range of materials to make a model of a plant.

2   When you choose your materials, think about the function (the job) of each part of the plant.

> You can work with a partner, but you both need to make your own model of a plant.

- **Roots:** You could make the roots from string or wool. This material will help the plant take in **water** from the soil where it is **growing**.

- **Stem:** You could make the stem from drinking straws. These are like the tiny tubes that transport water and nutrients around the plant.

- **Leaves:** Paper towels make good leaves. If you hold a paper towel up to the **light** you will see tiny holes. These are like the tiny holes in leaves that allow gases to move in and out of the plant.

- **Flowers:** You could make the flowers from coloured tissue paper, paper or fabric. Make them bright and colourful to attract insects.

3   Put your plant in a container or pot filled with sand or soil. Make sure your plant is standing up straight and does not fall over.

Parts of a flowering plant

See Student Book 3, pages 52–53

Home learning Main parts of flowering plants

Take home the model of the flowering plant that you made in class.

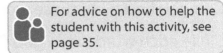

For advice on how to help the student with this activity, see page 35.

Show the people at home your model of a flowering plant.

■ Point out (**identify**) the flower, roots, stem and leaves.

■ Tell the people at home about the function of each part of the flowering plant.

■ **Explain** why you chose the materials for each part. Make sure you link your reasons to the function of each part.

Complete the table with the people at home.

Part of the plant	Material used	Reason for using the material
Roots		
Stem		
Leaves		
Flowers		

Give the people at home four sticky notes. Ask them to write the name of each part of the plant on a separate sticky note.

Ask them to stick the labels onto the correct parts of your model.

Did they get them all right?

What do plants need so they can grow?

See Student Book 3, pages 54–54

Home learning Scientific enquiry wordsearch

Can you find all of these words in the wordsearch?

Circle each word when you find it.

They are all words to do with scientific enquiry skills.

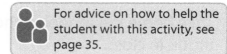

For advice on how to help the student with this activity, see page 35.

| predict | plan | fair test | observe |
| results | graph | table | measure |

g	o	i	n	t	t	a	b	l	e
g	r	a	p	h	a	p	b	a	k
v	g	w	l	t	p	l	a	n	x
p	u	g	v	y	k	h	l	l	z
f	t	m	e	a	s	u	r	e	j
a	l	j	x	s	h	o	m	q	e
f	r	e	s	u	l	t	s	o	l
v	o	b	s	e	r	v	e	y	s
q	w	p	r	e	d	i	c	t	q
f	a	i	r	t	e	s	t	f	w

When you find each word, **discuss** with an adult at home what you think it means.

What do plants need so they can grow?

See Student Book 3, pages 56–59

Class activity Plant investigation

✋ **Investigating plants**

Work with a partner.

Decide what you want to find out. You can choose to **investigate**:

■ whether plants need water to grow

or

■ whether plants need light to grow.

> If you would like to, you can repeat this activity for the investigation you did not choose this time.

✏️ Complete the sections below to help you plan your investigation.

What is your question?

_____

What do you predict will happen to the plant in your investigation?

_____

Why do you think this?

_____

Planning the investigation

I will need _____

_____.

I going to change _____.

I going to keep _____ the same.

I am going to **measure** _____.

Making observations

How will you make your observations accurate? _____

Which measuring device will you use? _____

How will you keep your results neat and tidy? _____

💬 Share your plan with the rest of the class.

3 Flowering Plants

39

What do plants need so they can grow?

See Student Book 3, pages 58–59

Home learning Photosynthesis competition

1 Work with two teams at home. Ask both teams to make as many words as they can using the letters from the word

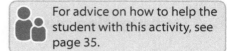

For advice on how to help the student with this activity, see page 35.

Photosynthesis

Ask someone from each team to write each word they have made onto a separate piece of paper.

Count the number of words that each team made. Record the number in the table.

Make sure the words are correct! Use a dictionary if you need to.

	Team A	Team B
Number of words		

2 Find two more words that begin with 'photo-'. You can use a dictionary for this too.

Write the words here.

How do plants take in water?

See Student Book 3, pages 60–61

Home learning Grow a bean seed

 Grow a bean seed in a jar.

For advice on how to help the student with this activity, see page 35.

> You will need: a glass jar, paper towels or blotting paper, an uncooked kidney bean or other bean and water.

Set up the growing jar as shown in the picture.

Place the jar on a windowsill where it will get lots of sunlight.

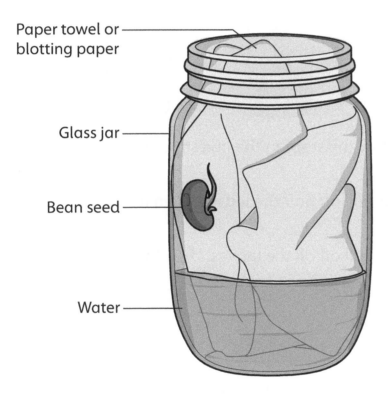

Paper towel or blotting paper

Glass jar

Bean seed

Water

Observe your bean seed every day for a few weeks.

 Draw your observations on a piece of paper every few days.

GOING FURTHER

> Why do the roots grow downwards?
>
> Why does the stem grow upwards?

How do plants take in water?

See Student Book 3, pages 62–63

Class activity Do leaves affect how a plant transports water?

✋ Does the number of leaves affect how water is transported?

You have already carried out an investigation using celery.

Now you are going to **investigate** whether the number of leaves on the celery affects how water is transported in the stem.

Work in a group of three or four.

> You will need three pieces of celery.

1 Take one piece of celery and follow the steps in the investigation on page 62 of your Student Book.

2 Take another piece of celery and pull off half the leaves. Then follow the same steps.

3 Take another piece of celery and pull off all the leaves. Then follow the same steps.

✏️ Predict what you think will happen.

_____

Check what happens to the celery over the next two days.

📏 Use a ruler to **measure** how far up the stem the dye has moved.

✏️ Use a table to record your observations.

✏️ Does the number of leaves affect the way the dye moves through the stem? Was your prediction correct?

_____

Healthy plants

See Student Book 3, pages 64–65

Home learning The power of roots

Roots can be very strong. They can even force their way through bricks and through road surfaces.

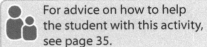

For advice on how to help the student with this activity, see page 35.

 Look for strong roots.

Look around the area where you live.

Try to find some roots that are growing like the examples in the photograph.

Have they caused any damage?

Draw the roots you find.

Healthy plants

See Student Book 3, pages 66–67

Home learning Tree survey

How many trees can you find?

Carry out a survey of all the trees you see on your way home or in the area where you live.

Keep a tally chart to record the number of trees that you find.

Ask the people at home if they know the names of the trees in your area.

Ask them if these trees have any special uses.

You can also use books or the Internet to help you **identify** the trees and their uses.

Use the table to record your results neatly.

For advice on how to help the student with this activity, see page 35.

Look at this tally to help you remember how to make a tally chart.

Tally	Number
					6

Name of tree	Tally	Number	Uses

Compare your findings with others in your class who live in different areas. **Discuss** any differences you find.

Not too hot and not too cold!

See Student Book 3, pages 68–69

Class activity Seedling investigation

✋ In what conditions do plants grow best?

You are going to **investigate** which combination of light and **temperature** helps plants to grow best.

Work in a group of three or four.

1 Half fill each pot with **compost**.

2 Carefully push a seed into each pot, until the seed is covered with the compost.

3 Water the pots and place one pot in each of the places shown in the table.

4 Check the pots every day. A plant should grow in each one.

> You will need: six pots, some compost, seeds and labels.

📏 **Measure** how tall the plant grows. Count the number of leaves. **Observe** whether it looks **healthy** or **unhealthy**.

✏️ Predict which plant you think will be the healthiest. _____

Use the table to record your results neatly.

Place	Height (cm)	Number of leaves	Observations
a Cool, dark place			
b Cool, light place			
c Warm, dark place			
d Warm, light place			
e Hot, dark place			
f Hot, light place			

✏️ My conclusions are _____

_____ .

Home learning Make your own greenhouse

 Design and make a greenhouse for a plant.

You will need to work with an adult.

Ask the adult to help you find materials to make your greenhouse.

Cling film and polythene are good materials for the top. Plastic food boxes or trays are good for the container.

For advice on how to help the student with this activity, see page 35.

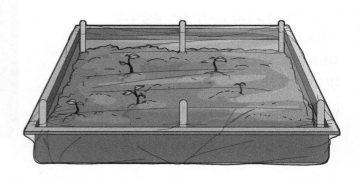

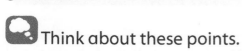 Think about these points.

■ What container will you use for the plant to grow in?

It will have to keep the soil and water in.

■ How big does the container need to be?

If you are growing seedlings the tray does not have to be very deep.

■ Does the greenhouse need to let light through?

If it does, what material will you use for the top?

■ Do you need to remove the top?

You might need to check the height of your plant or water it.

 Design your greenhouse and then build it.

What we have learned about flowering plants

See Student Book 3, pages 70–71

Home activity What I have learned...

Answer the questions to assess your understanding of this module.

For advice on how to help the student with this activity, see page 35.

1 Name the four parts of the plants you have learned about.

a _____

b _____

c _____

d _____

2 Talk about the function of each part of the plant.

Write the function of each part of the plant.

a _____

b _____

c _____

d _____

3 Write two things that roots do for a plant.

a _____

b _____

4 Is the sentence true or false? Circle your answer.

Trees are plants.   True   False

4 Introducing Forces

Extra support

Introduction

In this module students learn about forces, focusing on pushes and pulls. Students will be familiar with the idea that we cannot see forces but we can see the effects of forces all around us. They review their learning by stating examples of where they have seen a push or pull force. They are then encouraged to make a forcemeter to help them begin to understand that some forces are greater than others and that this can be measured. Students then explore how forces can change the shape of some objects and can make others move or stop. They investigate how the type of surface can affect how objects move across it – how fast and how far given the same force. They look at the soles of different types of shoes and how these affect the use of the shoe.

This module will help students to practise these scientific enquiry skills:

- observation – collecting evidence by looking and measuring (pages 50, 51, 52, 53, 54, 55, 56, 57, 58, 59, 60)

- planning – asking questions and planning how to seek answers (pages 52, 53, 54, 55, 56, 57, 59, 60)

- predicting – stating what they think will happen and then comparing this with what actually happens (page 57)

- recording – writing or drawing observations or stages in work (pages 50, 51, 52, 54, 55, 56, 58, 59, 60)

- making comparisons – comparing sets of evidence or data (pages 52, 54, 55, 57, 58, 59, 60)

- drawing conclusions – examining results to identify any patterns and/or to suggest explanations (pages 55, 57, 58).

Ways to help

Forces are a key concept in science but they are an extremely abstract concept, as forces cannot be seen. It is very important to point out everyday examples of forces – such as the many pushes and pulls we use – to make the topic more concrete. Encourage the student to identify forces as they do everyday activities such as opening and closing doors and playing games. Use examples in the kitchen of forces being used to change the shape of something (for example push and pull forces when kneading dough) and show how objects such as jewellery are made by stretching, bending or twisting metals. Play games to show that forces are needed to start and stop objects and that some surfaces have less friction than others.

Remember to encourage and help the student but try to let them find out as much as they can on their own.

Finally, help the students complete the 'What I have learned…' summary to test their understanding and recall.

Key words

direction	pull
force	push
forcemeter	stop
friction	weight
Newtons	

Scientific enquiry words

compare	pattern
draw	predict
fair test	present results
investigate	record
measure	table

 ## Helping with activities

The following guidance is intended to offer advice to the parent, or other adult at home, on how to help the student with each home learning activity.

Pushes and pulls at home (page 51)

To review learning, the student is asked to draw pictures of a push and a pull that they have seen or done at home. Support them in drawing arrows on the pictures to show the direction of the force.

Make a forcemeter (page 52)

Help the student to make a simple forcemeter using an elastic band attached to an object. Select some suitable low-mass objects, such as a pencil, a small toy car and a spoon. You can also help the student to make a simple scale using labels stuck to the ruler.

Make your own play dough (page 53)

You can use any kitchen equipment to mix the dough as only foodstuffs are used. Safety – do not allow the student to add the boiling water or to touch the dough until it has cooled down.

Encourage the student to squash, squeeze, pull and roll the dough to make different shapes. Talk about how pushes and pulls change its shape.

Bouncing ball investigation (page 55)

A soft foam ball is best for this activity, as it will not bounce too quickly or too far. The student should be able to tell you that throwing the ball uses a push force. Time the ball from when it leaves the student's hand to when it bounces back. Talk about how hard or gently the student needs to throw the ball to make it bounce back.

Surface investigation (page 58)

Encourage the student to push and pull objects across the surface without and then with an ice cube. They should see that the ice helps the objects move easily. You could use the forcemeter from the 'Make a forcemeter' activity to demonstrate that using the ice means less force is needed.

Which shoe? (page 59)

Look at the types of shoes and talk about when each type would be worn. Then read out the uses. Encourage the student to link each shoe to its use. Talk about friction and grip.

Rolling a ball (page 60)

You will need two tennis balls or soft foam balls and a tape measure or ruler. For each challenge, roll the ball three times – this introduces the concept that one try might not give a reliable result. You can talk about which try was the most accurate. If two or all three tries give the same result, this means you can rely on that result being reliable. This activity supports the idea that a force (in this case a push) results in an action and that the force can vary in intensity.

What I have learned... (page 61)

The review questions are intended to gauge what the student has learned in this module. Talk through the questions and answers and listen out for any misunderstandings.

Pushes and pulls

See Student Book 3, pages 74–75

Class activity Find the forces

Label all the **pushes** and **pulls** you can see in the picture.

Draw an arrow to show the **direction** of each **force**.

push	pull

Discuss your ideas. You might not agree on them all.

Pushes and pulls

See Student Book 3, pages 74–75

Home learning Pushes and pulls at home

1 Draw a picture of a push you have seen or done at home.

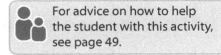

For advice on how to help the student with this activity, see page 49.

Draw an arrow on your picture to show the direction that the push is moving in.

2 Draw a picture of a pull you have seen or done at home.

Draw an arrow on your picture to show the direction that the pull is moving in.

Home learning Make a forcemeter

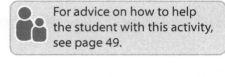

For advice on how to help the student with this activity, see page 49.

Make and use a **forcemeter**.

1 Use an elastic band to make your own forcemeter.

You will attach different objects to the elastic band.

This allows you to **measure** the force needed to lift each object. The longer the elastic band stretches, the bigger (greater) the force.

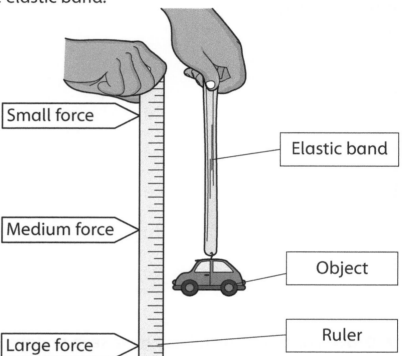

Small force

Elastic band

Medium force

Object

Large force

Ruler

A forcemeter has a scale to help us **measure** the force accurately. You can use a ruler to **measure** the length of the elastic band. Ask an adult at home to help you do this.

Make a scale for your forcemeter. You can use these terms:

■ 'large force' – when the elastic band stretches a lot

■ 'small force' – when the elastic band does not stretch very much

■ 'medium force' – in the middle between large force and small force.

2 Use your forcemeter to lift small objects in your home.

3 Look at the scale and observe the size of the force needed to lift each object.

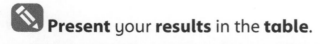**Present** your **results** in the **table**.

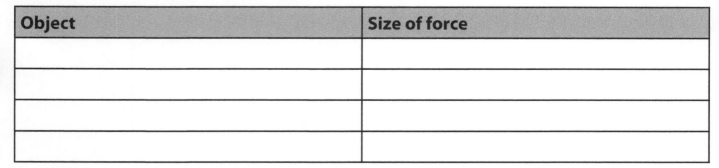

Object	Size of force

Making shapes with forces

See Student Book 3, pages 78–79

Home learning Make your own play dough

Make your own play dough.

You have learned how forces can change the shape of objects. Now you can **investigate** this by making and using your own play dough.

For advice on how to help the student with this activity, see page 49.

You will need:

- 2 cups plain flour (all purpose)
- ½ cup salt
- 2 tablespoons cream of tartar
- 2 tablespoons vegetable oil
- 1.5 cups boiling water
- food colouring (optional)
- few drops glycerine (optional: adds more shine!)

Method:

- Mix the flour, salt, cream of tartar and oil in a large mixing bowl.

- Add food colouring (optional) to the boiling water. Pour the coloured water into the dry ingredients a little at a time.

- Stir continuously until you have a sticky, combined dough.

- Add the glycerine (optional). Then allow the dough to cool down.

- Take the dough out of the bowl and knead it vigorously for a few minutes until all of the stickiness has gone. This is the most important part of the process, so continue until the dough is the perfect consistency.

You must ask an adult to help you make the dough.

The adult must add the boiling water and stir the dough. Do not do this yourself.

Squeeze, squash, stretch and roll the dough to change its shape.

Think about how each type of force changes the shape of the dough.

Making shapes with forces

See Student Book 3, pages 78–79

Class activity Use forces to change the shape of modelling clay

Use forces to change the shape of modelling clay.

This activity provides support for the first investigation on page 79 of your Student Book.

1 Use the modelling clay to make a pot.

2 Squash, stretch and squeeze the clay to make it into the shape that you want.

Compare your pot with another student's pot.

Are the pots the same?

If not, how do they look different?

Are the pots the same size?

How can you make sure that you both make a pot of the same size?

How can you **measure** this?

_____

Planning my investigation

I will need

_____ .

What am I going to change?

_____

What am I going to keep the same (to make this a **fair test**)? _____

_____

What am I going to **measure**?

_____

What am I going to do? _____

_____

Forces can stop things moving

See Student Book 3, pages 80–81

Home learning Bouncing ball investigation

 Use forces to control a ball.

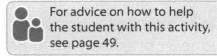

For advice on how to help the student with this activity, see page 49.

> You will need: a bouncing ball (a soft foam ball is best for this investigation) and a stop watch or a clock or watch with a second hand (to **measure** the time in seconds).

1 Find a safe place to **investigate** your bouncing ball. Ask an adult to help you with this.

2 Throw the ball at a wall.

Is this a push or a pull? _____

3 Can you throw the ball at the wall so that it bounces back?

4 Try to control how fast the ball bounces back.

- Throw the ball and ask your helper to start the clock or timer.

- Can you make the ball take 5, 4, 3, 2, 1 seconds to return to you?

- Make sure you always stand the same distance from the wall.

5 Ask other people at home to try this.

Can you find a **pattern** in your results?

How do you change the way you throw the ball to make it come back more quickly?

_____

_____

How do you change the way you throw the ball to make it come back more slowly?

_____

_____

Forces can stop things moving

See Student Book 3, pages 80–81

Class activity Measuring distance

This activity provides support for the toy car investigations on page 81 of your Student Book.

In your investigations you need to **measure** the distance that a toy car travels across the floor.

What piece of equipment will you use to **measure** the distance?

Which one of these is the best to use? Circle your answer.

30-centimetre ruler   tape measure   metre stick   steel ruler

Look at the picture of the ruler.

It is important to start measuring from the beginning of the scale.

Draw an arrow to show the beginning of the scale on this ruler.

When you are measuring the distance the toy car has travelled, you must **measure** to the same point on the car every time.

Choose the front or the back of the car and **measure** to that point every time. This helps to make it a **fair test**.

Practise **measuring** using a ruler. Use your ruler to **draw** lines of these lengths in the spaces below.

2 centimetres

5 centimetres

12 centimetres

Forces can stop things moving

See Student Book 3, pages 82–83

Class activity Vehicles travel further on some surfaces

✋ How well do things move on different surfaces? (See pages 82–83 of your Student Book.)

You are going to **investigate** how different surfaces affect the distance that a toy car travels.

Work with a partner to plan your investigation.

What do you **predict** will happen to the distance the car travels on different surfaces?

Predict what will happen

I think that _____

_____ .

My reason is that _____

_____ .

My investigation plan

I will need _____ .

What am I going to change? _____

What am I going to keep the same? _____

What am I going to **measure**? _____

What am I going to do? _____

_____

I will be careful of _____ .

My drawing of what I will set up

Forces can stop things moving

See Student Book 3, pages 82–83

Home learning Surface investigation

How does the ice skater move so quickly?

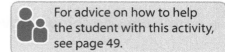 For advice on how to help the student with this activity, see page 49.

How easily do objects move on a surface?

1 Place some different small objects on a work surface or table.

> Ask permission from an adult before you do this.
>
> You may need to cover the surface with a waterproof covering or plastic sheet to protect it.

2 Push and pull the objects to move them across the surface. Is it easy to move them?

How hard do you have to push and pull the objects to make them move?

You could use the forcemeter you made to **measure** the force needed to move each object.

3 Now use an ice cube. Carefully place each object on the ice cube so that it is balanced.

4 Push and pull the objects to move them, as you did before. Do the objects move more easily now?

Draw a **table** to **record** your results.

....

Compare the movement of the objects with and without the ice cube.

How does the ice affect **friction** on the surface?

See Student Book 3, pages 84–85

Home learning Which shoe?

Look at the pictures of footwear. Look at the list of uses. Which footwear is best for each of the uses?

For advice on how to help the student with this activity, see page 49.

Draw a line to link each piece of footwear to its use.

Ice skating

Dancing

Walking

Ice climbing

Playing football

Running on a track

Running on mud

How have the shoes been designed to change the amount of friction?

Forces can affect speed and direction

See Student Book 3, pages 86–87

Home learning Rolling a ball

Work with an adult at home for this activity.

Find a space in front of a wall.

> For advice on how to help the student with this activity, see page 49.

Practice

Sit on the floor in front of the wall.

Roll the ball so it hits the wall and bounces away.

> You will need: two tennis balls or soft foam balls and a tape measure or ruler.

 Measure how far it bounces away.

The challenge

1 Who can get the ball closest to the wall?

Roll the ball so it **stops** as close to the wall as possible. Have three turns each.

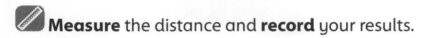 **Measure** the distance and **record** your results.

2 Who can make the ball stop 1 metre away from the wall?

Roll the ball so it bounces off the wall and stops as close to 1 metre away from the wall as possible. Have three turns each.

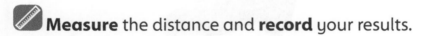 **Measure** the distance and **record** your results.

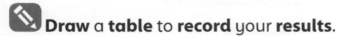 **Draw** a **table** to **record** your **results**.

💬 Think about what you had to change to get the ball as close as possible to the point you were aiming for.

What we have learned about introducing forces

See Student Book 3, pages 88–89

Home activity What I have learned...

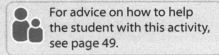

For advice on how to help the student with this activity, see page 49.

Complete the tasks and questions to assess and review your understanding of this module.

1 What force do you use for each of these activities? Write a word from the word bank each time.

   a Open a drawer _____

   b Kick a ball _____

   c Move a swing backwards _____

   d Throw a ball _____

   e Pick up a book _____

   | push   pull |

2 True or false? Circle your answer each time.

   a **Weight** is not a force.   True   False

   b Weight is **measured** in **Newtons**.   True   False

   c Forcemeters **measure** force.   True   False

   d You can make your own forcemeter.   True   False

   e Friction is a force.   True   False

   f Friction makes things move faster.   True   False

5 The Senses

Extra support

Introduction

In this module students will explore their senses of touch, taste, seeing, hearing and smell. There are various activities for students to try, requiring equipment and materials that should be readily available. The taste activities should be closely supervised to ensure that the student only samples the food you have prepared. Hearing and seeing activities may have to be approached with sensitivity if anyone has a visual or hearing impairment.

This module will help students to practise these scientific enquiry skills:

- observation – collecting evidence by looking and measuring (pages 64, 65, 66, 67, 68, 69, 70, 71, 72, 73, 74)

- planning – asking questions and planning how to seek answers (pages 65, 66)

- predicting – stating what they think will happen and then comparing this with what actually happens (page 67)

- recording – writing or drawing observations or stages in work (pages 64, 65, 67, 68, 69, 70, 72)

- making comparisons – comparing sets of evidence or data (pages 64, 65, 66, 68, 71, 72)

- drawing conclusions – examining results to identify any patterns and/or to suggest explanations (page 67).

Ways to help

Provide as many examples as you can of how humans use their senses to interact with their surroundings. Encourage the student to use only touch to identify objects, and then help them to explore the other senses – smell, taste, hearing and sight. Use mealtimes to ask about how they taste foods and help them to link smell and taste. They can also talk to you about optical illusions they have seen. Encourage them to find examples of smells in nature – for example, the scent of flowers. Having a vase of cut flowers at home or flowers in a pot or garden will provide opportunities for this.

Remember to encourage and help the student but try to let them find out as much as they can on their own.

Finally, help the students complete the 'What I have learned…' summary to test their understanding and recall.

Key words

carbohydrate	nose
ears	optical illusion
eyes	see/sight
fingerprints	senses
food	smell
hear/hearing	taste
materials	tongue

Scientific enquiry words

compare	identify
discuss	make
explore	think
group	

 ## Helping with activities

The following guidance is intended to offer advice to the parent, or other adult at home, on how to help the student with each home learning activity.

Family fingerprints (page 65)

Explain that everyone has unique fingerprints but that we share similar patterns. The method used to take fingerprints can be messy, so cover the table with newspaper or plastic sheeting. Have a bowl of soapy water ready to wash off the ink or paint.

Warm or cold? (page 66)

Safety – the warm water should be no hotter than warm to the touch. Make sure you cover the table and work away from electrical devices or sockets. Ensure that participants follow the instructions carefully or the activity probably will not work. Talk about how the skin does not always sense temperature correctly. Talk about how this can be dangerous: it is easy to get scalded when bathing and burned when cooking, for example.

Why should we chew our food? (page 68)

Use a small piece of any kind of bread. Make sure the student takes a long time to chew the bread.

Different smells (page 69)

Students should readily recall smells from perfume, flowers and cooking. You could provide some flowers or an air freshener. Encourage the student to talk about smells, especially the ones they would miss if they lost their sense of smell.

Flowers and their scents (page 70)

Remind the student why flowers have scents – to attract pollinating insects. Look at the photograph of the Brazilian flower and point out that scents may not always be pleasant for us. Take the student to a garden with lots of flowers or provide cut flowers. This activity will develop the student's communication skills.

How important is eyesight? (page 71)

Look at the photographs with the student and discuss how the visually impaired people are taking part in their sports. The investigation demonstrates how much we rely on seeing without thinking about it. Ensure that the student is sitting securely before you blindfold them. Give them directions to help them to eat their meal but do not physically help or the experience will be diminished.

Using Morse code (page 74)

The student has used some Morse code to tap out messages at school. This activity introduces the codes for more letters. Use the table to practise tapping out letters. Use a sharp tap with the knuckles for a dot, and a softer tap with the palm of the hand for a dash. Look at the chart while the student taps out the words.

What I have learned… (page 75)

Allow the student to work on this activity independently or provide support if required. You could help them find the first word to check they understand how a wordsearch works.

Class activity A touch display

✋ Sorting materials

Work with a partner.

1 Your teacher will give you different **materials**.

2 Carefully feel the materials with your fingers. Sometimes it helps us to **think** well if we close our **eyes**. Are the materials rough or smooth?

💬 **Discuss** with your partner how the materials feel. Decide together whether each material is rough or smooth.

If it is difficult to decide about some materials, **make** a **group** of materials that are neither rough nor smooth.

3 Sort them into **groups** of rough, smooth and neither rough nor smooth materials.

4 Use your materials to **make** a big class display.

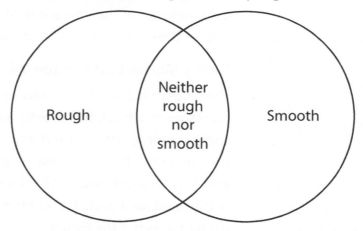

You will use a Venn diagram like the one in the picture, but larger. You might have seen one of these before. In this one there is a section for the rough materials and a section for the smooth materials. The section in the centre where the circles overlap is for the neither rough nor smooth materials.

5 Choose a small piece of each material or cut the material.

6 Stick the materials in the correct place on your Venn diagram.

Touch

Home learning Family fingerprints

Take the **fingerprints** of the people at home.

Explore the different fingertip patterns by taking fingerprints.

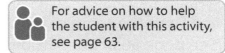For advice on how to help the student with this activity, see page 63.

Method

You will need an inkpad. Make sure the ink is easy to wash off. You can **make** your own inkpad with a sponge and some watercolour paint or poster paint. Let the paint soak into the sponge.

1 Place the edge of your fingertip next to the nail, on the inkpad. Carefully and slowly roll your fingertip across the inkpad.

2 Roll your fingertip onto a piece of paper in the same way as you did on the inkpad. Use matt paper as shiny paper can make the fingerprint smudge.

You might need some practice to get just the right amount of ink on your finger. Too much ink will make a splodge. Too little ink will make a very faint fingerprint that is hard to see.

Collect the fingerprints of all your family members. Write the name of each person next to their fingerprint.

Simple loops   Double loops   Whorl   Arch

Use these pictures to **identify** the fingerprint patterns of the people at home.

Compare your fingerprints. What similarities and differences can you **see**?

Touch

Home learning Warm or cold?

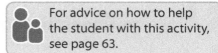 Does the sense of touch sometimes trick us?
(See page 96 of your Student Book.)

For advice on how to help the student with this activity, see page 63.

Demonstrate the investigation you did at school to the people at home.

1 Prepare three containers of water. You need warm water, room temperature water and cold water.

Ask an adult to help you prepare the containers of water. Make sure the water is warm, not hot.

Warm water   Room temperature   Cold water
                          water

2 Choose two people to do this investigation.

3 Tell the first person to place their right hand in the container of cold water and their left hand in the warm water.

4 Make sure they leave their hands in the water for one minute.

5 Then tell them to put both their hands in the room temperature water.

Ask them what they felt. Did the room temperature water feel the same or different for each hand?

6 Now ask the second person to do the same thing.

Did both people feel the same thing?

Ask more people to do the investigation. Try it yourself.

Discuss what everyone felt when they did the investigation.

See Student Book 3, pages 98–99

Class activity Taste and colour

✋ What does it **taste** like?

Your teacher will give you four cups of juice labelled 1 to 4. Each drink is a different colour. Can you tell from the colour what the drink will taste like?

1 Look at each drink and **think** about how it will taste.

✏️ Record your predictions in the table.

Drink	Prediction	Taste
1		
2		
3		
4		

2 Use your straw to taste each drink. Only use your straw.

✏️ Record how each drink tasted in the table.

Use the words in the word bank to help you.

salty   sour   sweet   bitter   savoury

💬 How accurate were your predictions? Can you trust your eyes to tell you what things will taste like?

Home learning   Why should we chew our food?

✋ Does chewing change the taste of **food**?

For advice on how to help the student with this activity, see page 63.

1 Place a small piece of bread on your **tongue**.

What does it taste like?

_____

2 Now chew the piece of bread for two minutes.

What does it taste like now?

_____

3 You can swallow the bread or spit it out into a paper towel.

GOING FURTHER

Bread is a **carbohydrate**. It contains starch. When you first put the bread in your mouth it tastes different from when you have chewed it. This is because carbohydrates are made of very big units. When you chew the bread it mixes with your saliva. A substance in the saliva called amylase breaks the units up. The smaller units are sugar. This is why the bread tastes sweeter after you have chewed it.

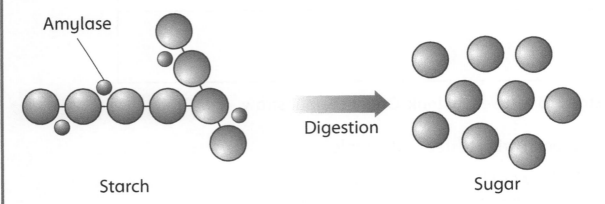

Amylase

Starch

Digestion

Sugar

✏️ Can you list any other reasons for chewing your food?

_____

_____

68

Smell

See Student Book 3, pages 102–105

Home learning Different smells

How does our sense of **smell** give us pleasure?
List three ways.

For advice on how to help
the student with this activity,
see page 63.

1 _____

2 _____

3 _____

How does our sense of smell warn us of danger? List three ways.

1 _____

2 _____

3 _____

What smells would you miss if you did not have a sense of smell? List three
smells you would miss.

1 _____

2 _____

3 _____

Ask people in your family what smells would miss if they did not have
a sense of smell.

Write their answers.

1 _____

2 _____

3 _____

Smell

See Student Book 3, pages 102–105

Home learning Flowers and their scents

Flowers give off scents to attract insects.

The photograph shows a flower in Brazil that smells like rotten meat to attract flies.

 What do the flowers near your home smell like?

Work with an adult to **explore** the scents of some different flowers. Complete the table.

Name of flower	Description of the smell

The large carrion flower smells of rotten meat to attract flies.

For advice on how to help the student with this activity, see page 63.

GOING FURTHER

Look at the picture. It shows how smells travel to our **nose** and brain. Label the process using the words in the word bank.

nostril	nose	flower

Explain to someone at home how you can smell the scent of a flower that is many metres away.

Nerves that detect smell

Sight

See Student Book 3, pages 106–107

Home learning How important is eyesight?

Look at the photographs. These sportspeople have a visual impairment. How are they taking part in their sports?

For advice on how to help the student with this activity, see page 63.

How easy is it to eat a meal without **sight**?

Try this when you are about to eat a meal with your family at home.

1 Sit in front of your meal.

2 Ask an adult to blindfold you. Stay in your seat.

3 Try to eat your meal with the blindfold on.

4 Your family can guide you but they must not feed you.

While you are eating, tell your family how the food tastes.

When you have finished your meal, tell your family how it felt to try to eat without sight.

What problems did you have?

Ask other people at home to eat a meal wearing a blindfold.

Compare the problems you all had.

5 The Senses

71

Class activity Tricking your eyes

Work with a partner.

✏️ Look at the picture.

Can you see any circles?   Yes   No

What colour are they?   _____

💬 **Discuss** with your partner what you can see.

✏️ Look at the two sets of lines. Which horizontal line is longer, line A or line B? _____

A

B

📏 Use a ruler to measure the length of the horizontal lines.

Top line _____ centimetres.

Bottom line _____ centimetres.

💬 Did your eyes trick you?

Visual tricks like these are called **optical illusions**.

See Student Book 3, pages 110–111

Class activity Lip reading

💬 Work with a partner.

Sit facing each other.

Student A: You will work without using your sense of **hearing**. Cover your **ears** with your hands.

Student B: Say a word from the word bank. Say the word a few times.

taste	touch
smell	hear
sight	sense

Student A: Look at your partner's lips as they say the word.

Can you tell which word your partner is saying?

You will have to use your scientific observation skills to do this.

What shapes do their lips make?

Can you see the position of their tongue?

Then swap over so you both have a turn at lip reading without using your sense of hearing.

💬 Repeat the activity. This time say a short sentence. Use one of the science sentences below.

> We breathe air through holes called nostrils.

> We use our tongues to taste things.

> Sometimes our eyes are tricked.

💭 People who cannot **hear** learn to lip read very skilfully.

People who work in noisy places also learn to lip read.

Did you find lip reading easy or difficult?

✏️ I found lip reading _____ .

Hearing

See Student Book 3, pages 112–113

Home learning Using Morse code

You have used a simple Morse code to tap out messages at school.

For advice on how to help the student with this activity, see page 63.

Now you can develop this skill further.

The letters you can use are shown in the tables.

Letter of the alphabet	Morse code dots and dashes
A	. –
E	.
I	. .
O	– – –
C	– . – .
N	– .
M	– –
T	–

Letter of the alphabet	Morse code dots and dashes
H	. . . .
S	. . .
G	– – .
L	. – . .
R	. – .
U	. . –
Y	– . – –

1 Teach the people at home to tap out the letters. Use the palm of your hand for a dash and your knuckles or fingertips for a dot.

2 Show the tables to the people you are working with.

3 Now tap out one of the words in the word bank. They are words for the **senses** and the sense organs.

Did the people at home work out the word?

4 Now ask them to tap out a word for you.

Did you work it out?

touch	taste	smell
ears	fingers	tongue
nose	sight	eyes

 Which sense is missing from the word bank? _____

What we have learned about the senses

See Student Book 3, pages 114–115

Home activity What I have learned...

This activity will help you remember what you have learned in this module.

For advice on how to help the student with this activity, see page 63.

✏️ Find 12 words used in this module in the wordsearch. Circle the words when you find them.

t	o	u	c	h	s	m	a	p	d
v	c	t	j	r	t	a	s	t	e
u	s	i	g	h	t	l	i	m	f
f	p	h	e	a	r	i	n	g	e
y	n	o	s	t	r	i	l	s	n
e	a	r	s	w	t	n	o	s	e
l	k	i	x	s	m	e	l	l	h
h	e	y	e	s	g	q	j	n	b
s	b	z	f	i	n	g	e	r	s
t	o	n	g	u	e	b	e	o	K
y	m	l	o	x	s	k	i	n	v

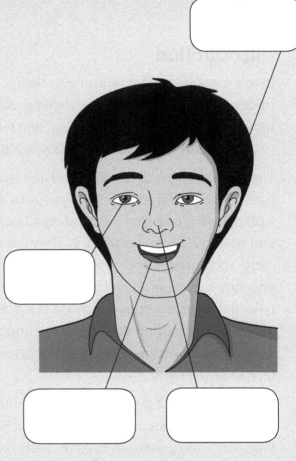

✏️ Use some of the words you have found to label the diagrams of a face and hand.

Write the name of the part of the body and the sense in each box.

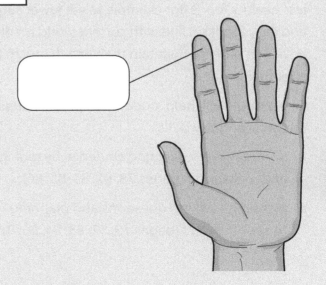

6 Keeping Healthy

Extra support

Introduction

In this module students will review their understanding of life processes (eating, drinking, moving, reproducing and growing) and relate this to ways that animals and plants stay healthy.

There is a strong focus on the food and water that humans need and students will carry out activities to explore the importance of eating a healthy diet and having clean water to drink. They will study the major food groups of carbohydrates, fats, proteins, vitamins and minerals and the need to eat a balance of these groups. Emphasis is placed on healthy choices. They also look at the importance of exercise and investigate how this affects our heart rate and breathing.

Some examples of diseases caused by nutrient deficiencies are discussed in the Student Book. Students are encouraged to eat healthy rather than less healthy foods (for example to eat fewer sugary and fatty foods) in line with current world medical advice, in order to maintain their health and to look after their teeth.

This module will help students to practise these scientific enquiry skills:

- observation – collecting evidence by looking and measuring (pages 78, 82, 83, 85, 86)
- planning – asking questions and planning how to seek answers (pages 79, 82, 83, 84, 86, 87)
- predicting – stating what they think will happen and then comparing this with what actually happens (page 82, 84)
- recording – writing or drawing observations or stages in work (pages 80, 81, 82, 83, 84, 85, 86, 87)
- making comparisons – comparing sets of evidence or data (pages 78, 80, 85, 86, 87, 88)
- drawing conclusions – examining results to identify any patterns and/or to suggest explanations (pages 78, 85, 86, 87, 88).

Ways to help

Use foods at home to help the student to understand the link between what they eat and their health. Point out the energy content of foods and the other nutrients they contain. Food labels will help you. Also explain the dangers of over- and undereating: eating too much food can be unhealthy, but there are many people who cannot get enough food. Work with the student and the rest of the family to discuss a healthy diet and to plan meals. Also reinforce the importance of water and exercise for a healthy lifestyle.

Remember to encourage and help the student but try to let them find out as much as they can on their own.

Finally, help the students complete the 'What I have learned…' summary to test their understanding and recall.

Helping with activities

The following guidance is intended to offer advice to the parent, or other adult at home, on how to help the student with each home learning activity.

Designing a poster (page 79)

The student's poster will show their ideas about helping people in the world who do not have enough food. Encourage the student to use the key words and if they want to add pictures try to provide newspaper or magazine cuttings.

Healthy eating (page 80)

Help the student to think about the various foods they could choose. Encourage them to include a range of different food groups to make a healthy, balanced meal. Health advice is that carbohydrates that release energy slowly (such as wholegrain cereals) are better than sugary foods.

Find the hidden words (page 81)

Wordsearches are a good way to help the student practise key words in science and to help with general literacy. Once they have found each word, ask them to say it out loud and tell you what it means. The missing food group is carbohydrates.

Looking after water (page 82)

Help the student to select a tap to test and to find a measuring jug. Make sure the tap is dripping slowly, otherwise the jug will overflow before the hour is over. You may need to help the student with the calculations. Multiply the volume collected in one hour to calculate the volume wasted in a day. Use a calculator to make this easy. Multiply the daily volume by 7 to find the volume wasted in a week. Then multiply the weekly volume by 52 to find the volume wasted in a year.

Food for a long walk (page 84)

The student plans a meal for people going on a long walk. Encourage the student to think about foods that are rich in energy. These are foods that contain a lot of energy per gram. This means the walkers will have good energy sources that are not heavy to carry. Walkers also need plenty of water. Generally, humans should drink 2 to 3 litres per day, but walkers may need at least 3 to 4 litres depending on the weather.

Favourite foods (page 87)

Answer the student's questions about your favourite foods and some eating habits. Help the student to record their findings in the table. The idea is not to talk about 'good' and 'bad' foods but to help the student to appreciate that too many sugary and fatty foods can be harmful to health.

Our teeth (page 88)

Encourage the student to look at their teeth and to discuss with you the different types of teeth. Help the student to understand that we need different types of teeth for different jobs.

What I have learned... (page 89)

Allow the student to ask you questions about what they have been studying at school. They can add their own questions if they wish. This activity reinforces their learning and lets them share their understanding and knowledge with others.

The life processes

Class activity How much should we eat?

The first table shows the energy contained in each main **food** group. Calories often appear on food labels, so both joules and calories are shown.

Food group	Energy in every gram	
	Kilojoules	Kilocalories
Fat	37	9
Protein	17	4
Carbohydrates	17	4
Fibre	8	2

Compare the energy in the food groups in the table.

Which food group contains the most energy in every gram? _____

Which food group contains the least energy in every gram? _____

Age in years	Average energy needed every day in kilojoules	
	Male	Female
0–1	2 000	2 000
2–4	6 000	5 000
5–10	7 000	6 000
11–15	11 000	8 000
16–25	12 000	9 000
26–45	11 000	10 000
46–65	10 000	9 000
66–80	8 000	7 000

Look at the second table. Why do young adults need more energy than older adults?

Why do 4-year-old children need more energy than babies?

Why do 15-year-old boys need more energy than 15-year-old girls?

The life processes

See Student Book 3, pages 118–119

Home learning Designing a poster

You are going to **design** a poster to tell people that some people in the world do not have enough food. You can do some **research** using the Internet or books.

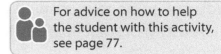

For advice on how to help the student with this activity, see page 77.

Explain why people need food and **water**.

Use the key words in the word bank.

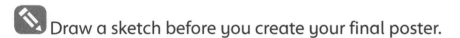

food   nutrition   healthy   water   help   give

Planning

In science, it is important to plan your work before you start.

- How will you make your poster bright and interesting?
- How will you make sure the message is clear and stands out?
- Will you add pictures to help you to explain the problem?
- How will you make sure people can read the key words?
- How will you suggest that people can help?

Draw a sketch before you create your final poster.

Diet and exercise

See Student Book 3, pages 120–121

Home learning Healthy eating

You are going to work with people at home to **design** a **healthy** breakfast. Here is a list of foods you can choose from.

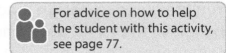

For advice on how to help the student with this activity, see page 77.

Beans	Banana	Bread	Eggs	Cheese	Chicken
Pancakes	Apples	Rice	Fried potatoes	Oranges	Muesli
Oatmeal	Nuts	Chickpeas	Fish	Tomatoes	Squash
Yoghurt	Cornflakes	Okra	Onions	Pain au chocolat	

Look at the picture of a balanced food plate. Remember that if we eat the correct amount of foods from each group, we have a balanced **diet**.

Fruits and vegetables = 33%

Carbohydrates = 33%

Dairy products = 15%

Fats and sugary food = 7%

Protein = 12%

Draw a large circle for the plate. Choose some foods for a healthy breakfast. Draw the foods or write their names on your plate.

Think about each food carefully. Do not just choose your favourite foods. **Think** about making a healthy, balanced meal.

GOING FURTHER

Can you estimate how many kilojoules the meal will contain? Use food labels or the Internet to help you with this.

 Write the estimate next to your plate.

Diet and exercise

See Student Book 3, pages 122–123

Home learning Find the hidden words

Find these words in the wordsearch.

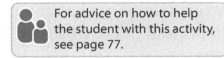

For advice on how to help the student with this activity, see page 77.

| vitamins | minerals | proteins | fats | scurvy |
| rickets | diet | nutrition | health | |

The words go across → and down ↓.

✎ Circle the words when you find them and write them in the box.

d	i	e	t	x	n	v	m	h	z	c
q	b	i	f	w	u	d	w	v	r	k
u	y	n	v	i	t	a	m	i	n	s
l	d	s	n	x	r	z	i	g	o	a
t	b	p	y	v	i	m	n	p	w	h
q	f	r	e	y	t	p	e	d	d	e
k	b	o	r	n	i	x	r	n	i	a
f	a	t	s	l	o	c	a	u	g	l
n	p	e	d	s	n	u	l	v	c	t
j	r	i	c	k	e	t	s	o	w	h
e	i	n	x	c	i	v	z	t	y	g
i	c	s	h	o	s	c	u	r	v	y

Which food group is missing from the list? _____

GOING FURTHER

You can use the Internet to help you with this question.

✎ How would you know if a person had scurvy?

_____

Home learning Looking after water

Water is important. Without water life cannot exist. It is important not to waste water.

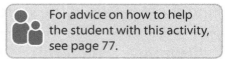
For advice on how to help the student with this activity, see page 77.

 How much water is wasted from a dripping tap?

Ask an adult at home to help you.

1 Turn a tap on slightly so it drips water. Try to make it drip so when one drip hits the sink another drip starts falling from the tap.

2 Place a measuring jug beneath the tap. Make sure the jug collects the water.

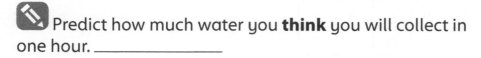 Predict how much water you **think** you will collect in one hour. _____

Remember to include the units.

3 Leave the tap dripping for one hour, then turn it off.

 Measure how much water you have collected in one hour.

 The volume of water collected in one hour is _____ .

Remember to include the units.

4 Work out how much water that dripping tap will waste in one day.

Remember, there are 24 hours in 1 day. Multiply the volume collected in 1 hour by 24.

 The volume of water wasted in one day will be _____ .

Remember to include the units.

 Can you work out how much water will be wasted in one week? Can you work out how much will be wasted in one year?

Diet and exercise

See Student Book 3, pages 124–125

Class activity Filtering water

It is not always easy for people to get clean water. Drinking dirty water can cause serious diseases. It is important to be able to treat water to make it cleaner. One way of doing this is to filter the water.

✋ Filter dirty water.

1 Set up your filtering apparatus.

2 Your teacher will give you some dirty water. **Observe** it carefully.

✏️ Write what it looks like.

3 Pour the dirty water through your filter.

⏱️ How long does it take to go through the filter?

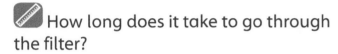

This is a smaller version of how water is cleaned for towns and cities. Very large filters use sand instead of cotton wool.

— Funnel

— Cotton wool

✏️ What does the water look like after it has been through the filter?

What does the cotton wool in the filter funnel look like?

Where has some of the dirt gone?

Warning! The water you have filtered might look clean, but it is not safe. Bacteria and viruses will pass through the filter.

GOING FURTHER

Design a way to make water cleaner using sand. Draw your ideas on a piece of paper.

Diet and exercise

See Student Book 3, pages 126–127

Home learning Food for a long walk

You are going to plan a meal for some people to take on a long walk.

Walking for a long time uses up a lot of energy.

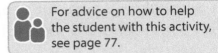

For advice on how to help the student with this activity, see page 77.

Plan a lunch that the people can carry and eat during their walk.

List all the foods you advise them to take.

_____   _____

_____   _____

_____   _____

Why did you select these foods?

_____

_____

How much water should the people take? Why?

_____

_____

Diet and exercise

See Student Book 3, pages 128–129

Class activity Energy for different activities

The amount of energy we need from our food depends on our age, whether we are male or female, our lifestyle and the job we do.

This person uses up a lot of energy every hour.

✏️ Look at the table and **compare** the energy used by each activity and sport. Then answer the questions.

Activity	Energy used every hour in kilojoules
Dancing	80
Gardening	60
Working at a desk	15
Working on a building site	90
Sport	Energy used every hour in kilojoules
Rowing	150
Cycling in a race	250
Playing football	120

Which of the activities needs the most energy every hour? _____

Why do you **think** this activity needs so much energy?

_____

Which of the sports needs the most energy every hour? _____

Why do you **think** this sport needs so much energy?

_____

💬 How much energy do you **think** you are using every hour to do your homework?

Class activity Heart rates

The number of times your heart beats every minute is called your heart rate. You can measure your heart rate using your pulse.

You can feel your pulse where you can feel a large blood vessel through your skin.

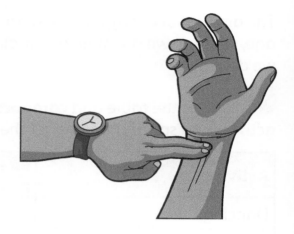

 Look at the picture. Practise feeling your pulse.

Count your heart beats for 30 seconds. Double this to work out your heart rate for one minute.

How does **exercise** change your heart rate?

1 Plan an investigation to **explore** how exercise changes your heart rate.

2 Carry out your investigation. Record your heart rates in the table.

Exercise	Heart rate (heart beats per minute)
Resting	
Walking slowly for a minute	
Walking quickly for a minute	
Running for a minute	

Remember:

- Take your pulse before you start exercising. This is your resting heart rate.

- You must rest after each exercise to let your heart rate recover.

Explain what happened to your pulse rate during exercise.

_____

Why did this happen?

_____

Home learning Favourite foods

 Are people's favourite foods healthy?

1 Choose three people to help you with the survey.

2 **Ask** each person the three questions and complete the table.

3 **Ask** yourself the questions as well and write your answers in the table.

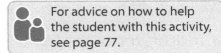 For advice on how to help the student with this activity, see page 77.

1 Name your five favourite foods.

2 How often do you eat sweets, cakes or chocolate in a day?

3 How often do you eat fruits and vegetables in a day?

Person's name	1 Five favourite foods	2 Number of times they eat sweets, cake or chocolate	3 Number of times they eat fruits and vegetables

Which favourite foods are the healthiest? _____

_____

Who eats the most carbohydrates? _____

Who eats the most fruits and vegetables? _____

Damaging foods

See Student Book 3, pages 132–133

Home learning Our teeth

We have different types of teeth to help us to eat our food.

- Incisors are at the front. They have sharp edges like scissors. They help us to bite food into smaller pieces.

- Canines are near the front. They are pointed. They help us to grip and tear our food.

- Molars are at the back. They are flat. They help us to grind and chew our food.

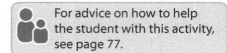

For advice on how to help the student with this activity, see page 77.

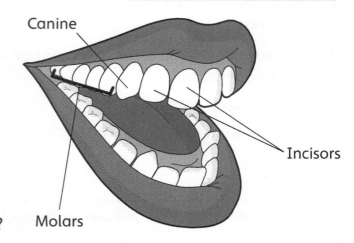

Canine

Incisors

Molars

  Lions have large canine teeth. Why?

Cows have large molars. Why? _____

Look at your own teeth.

You may still have your milk teeth. These will fall out to be replaced with adult teeth. The adult teeth are not replaced again so you will need to look after them!

Write three ways you can look after your teeth.

1 _____

2 _____

3 _____

Name three foods that you would not be able to eat easily if you did not have teeth.

1 _____   2 _____   3 _____

88

See Student Book 3, pages 134–135

Home activity What I have learned...

This activity will help you **think** about what you have learned in this module.

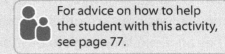

For advice on how to help the student with this activity, see page 77.

You are going to lead a family quiz.

1 Split the people at home into two teams called team A and team B. Try to have two or more people in each team.

2 **Ask** the first question to team A. If they say the correct answer, give them 2 points. If they say the wrong answer, **ask** team B the same question for 1 point. Record the scores.

3 If neither team says the correct answer, tell everyone the answer.

4 After you have asked all the questions, check how many points each team has. The team with the most points is the winner.

Here are the questions. You have learned the answers at school. If you need help to remember the answers, look at this module in your Student Book or use other books or the Internet.

What is **nutrition**?	Name one way that exercise helps to keep us healthy.	Name three food groups.	Why is insulin vital for our health?
Name two **life processes**.	What do our lungs do?	To keep healthy, we need to eat some of all of the food groups. What is this diet called?	What decays teeth?

 Which team won? _____

Quiz yourself

These quiz questions and activities are intended to encourage the student to reflect on their learning and to reinforce their developing knowledge about scientific concepts in a fun way. They are flexible enough to be individual, pair or group activities. Teachers and parents can use the quizzes in a number of ways.

- Questions can be selected from this section to supplement work carried out during each module, to act as extra tasks and support for individuals, groups and whole classes. In this way they can aid differentiation.

- Students can tackle the relevant questions at the end of each module to review learning and supplement the 'What we have learned…' sections.

- Students can undertake questions at the end of a series of modules or even at the end of the year to review learning. The questions could be set in batches over a series of lessons or even taken as a small timed test – although this is not their main purpose.

After each question the students can fill in the self-review circle to show how confident they are with that task. At Stage 3, students can draw a smiley face or put a sticker in the circle, or write a score from 1 to 5, with 5 being very confident. If a student reports that they are not confident in a certain area, the teacher can provide remediation.

1 Life Processes

1   Look at the pictures of animals. Decide which group each animal belongs to.

Draw a line from each animal to the correct circle.

Birds

Fish

Mammals

Self review How do I feel about this question?

2 Complete the table. Answer the questions about the object in each picture.

	Living or non-living?				
	Does it move?	Does it grow?	Does it breathe?	Does it eat?	Is it living or non-living?
(rabbit)					
(lamp)					
(frog)					
(saucepan)					

Self review How do I feel about this question? ◯

2 Materials

3 Follow the curly lines to match each material to one of its uses.

Write one property of each material in the box.

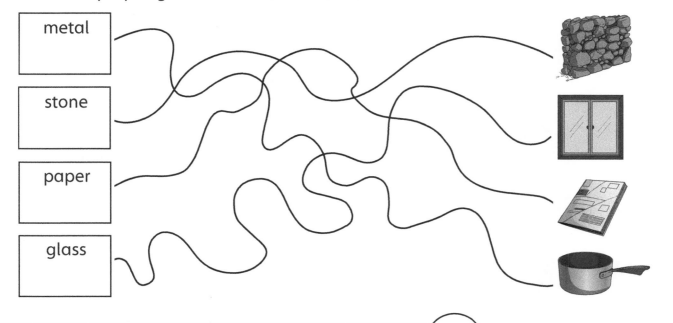

metal

stone

paper

glass

Self review How do I feel about this question? ◯

4 Look at the picture of an electromagnet hanging from a crane. Help the driver to sort the objects into magnetic and non-magnetic materials. Draw the magnetic objects on the electromagnet. Draw the non-magnetic objects in the wheelbarrow.

3 Flowering Plants

5 Look at the diagram of a plant.

a Label the plant using the words from the word bank.

b Read the information about the different parts of a plant. Draw a line from each label to the correct box.

1	This supports the plant and transports food and water around it.
2	These use the energy from sunlight to make food.
3	These keep the plant anchored in the soil. Some have hairs that help the plant to get water.
4	These often have a nice smell and colour to attract insects. This is also the place where seeds are produced.

stem   leaf   flower   roots

6 a Follow the maze and help the seedling to find what it needs to grow.

b What other things does the
seedling need for it to grow?

_____

_____

Self review How do I feel about this question?

4 Introducing Forces

7 Which force is being used in each picture? Write the correct word under each picture.

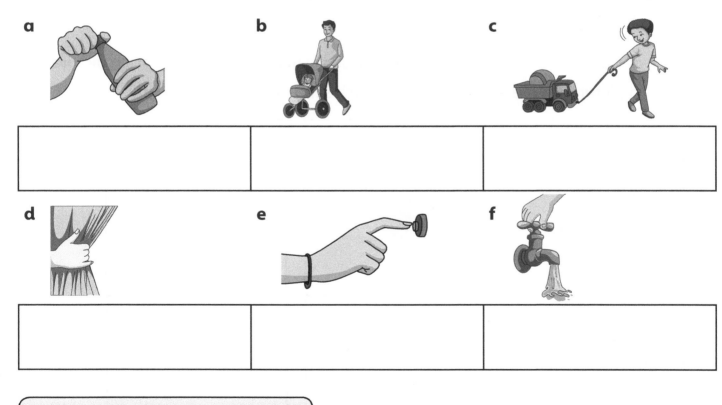

a

b

c

d

e

f

| push | pull | twist |

Self review How do I feel about this question?

8 Solve the clues for key words from this module. Write each key word in the correct place in the crossword.

Across

2 This force helps you to move an object towards you.

3 We use these to show the direction of forces.

5 This force gives you grip on your shoes.

6 This device measures force.

Down

1 This force pulls objects towards the centre of the Earth.

2 This force helps you to move an object away from you.

4 You use this to make objects move or change shape.

Self review How do I feel about this question?

5 The Senses

9 Look at the pictures. Write the name of the sense in the space above each picture.

Draw a picture of where you have used each sense. One is done for you.

hearing		
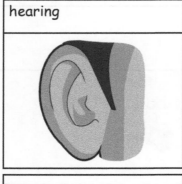	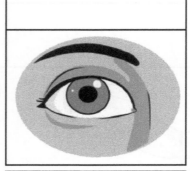	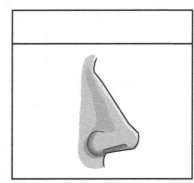

Self review How do I feel about this question?

10 a Look at the pictures. Circle or write your answer to the questions.

1	2	3
Is side A longer than side B? Yes No	Are the circles in a straight line? Yes No	Which is longer, line C or line D? _____

b Now use a ruler to check your answers.

1 Measure side A and side B. Write down the measurements. Side A: _____ Side B: _____	2 Line up your ruler with the circles. Are the circles in a straight line? Yes No	3 Measure line C and line D. Write down the measurements. Line C: _____ Line D: _____

c Which of your senses was tricked? _____

Self review How do I feel about this question?

11 Can you find in the wordsearch all of the words listed below? Circle each word when you find it. They are all words to do with keeping healthy.

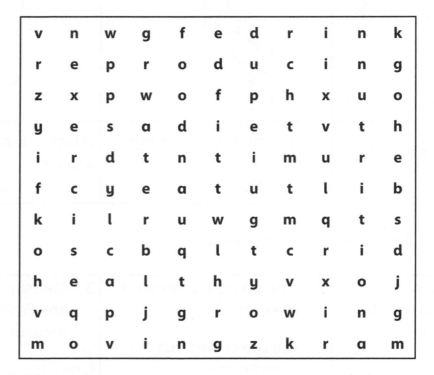

v	n	w	g	f	e	d	r	i	n	k
r	e	p	r	o	d	u	c	i	n	g
z	x	p	w	o	f	p	h	x	u	o
y	e	s	a	d	i	e	t	v	t	h
i	r	d	t	n	t	i	m	u	r	e
f	c	y	e	a	t	u	t	l	i	b
k	i	l	r	u	w	g	m	q	t	s
o	s	c	b	q	l	t	c	r	i	d
h	e	a	l	t	h	y	v	x	o	j
v	q	p	j	g	r	o	w	i	n	g
m	o	v	i	n	g	z	k	r	a	m

nutrition	drink
exercise	growing
diet	moving
healthy	reproducing
food	water
eat	

Self review How do I feel about this question?